Our Universe

Stars

by Margaret J. Goldstein

Lerner Publications Company • Minneapolis

Lerner Publications Company
A division of Lerner Publishing Group
241 First Avenue North
Minneapolis, MN 55401 USA

Website address: www.lernerbooks.com

Words in **bold type** are explained in a glossary on page 30.

Library of Congress Cataloging-in-Publication Data

Goldstein, Margaret J.
    Stars / by Margaret J. Goldstein.
      p.   cm. — (Our universe)
    Includes index.
    Summary: An introduction to the nature of stars, discussing their composition, size, color, formation, life cycle, constellations, clusters, and galaxies, as well as their study from Earth.
    ISBN: 0-8225-4646-9 (lib. bdg. : alk. paper)
    1. Stars—Juvenile literature. [1. Stars.] I. Title. II. Series.
QB801.7 .G65 2003
523.8—dc21   2002006837

Manufactured in the United States of America
1 2 3 4 5 6 – JR – 08 07 06 05 04 03

The photographs in this book are reproduced with permission from: © John Sanford, pp. 3, 17, 27. © Greg Vaughn/Tom Stack & Associates, pp. 6, 23; NASA, pp. 7, 9, 11, 15, 21. © Julian Baum/Photo Researchers, p. 13; © Jay M. Pasachoff/Visuals Unlimited, p. 19; © Science VU/Visuals Unlimited, p. 20; © Tsado/NASA/Tom Stack & Associates, p. 25; © NOAA/Tsado/Tom Stack & Associates, p. 26.

Cover © Jay M. Pasachoff/Visuals Unlimited.

Look up in the sky on a clear night. What do you see? Shining stars are twinkling in the dark sky. What are stars?

Stars are giant, glowing balls of gases. They send light and heat far out into space. There are trillions of stars in space. They are all different sizes.

From Earth, most stars look like tiny lights. Stars look small because they are very far away. But most are much bigger than Earth or any other planet.

star

EARTH AND STARS OF DIFFERENT SIZES

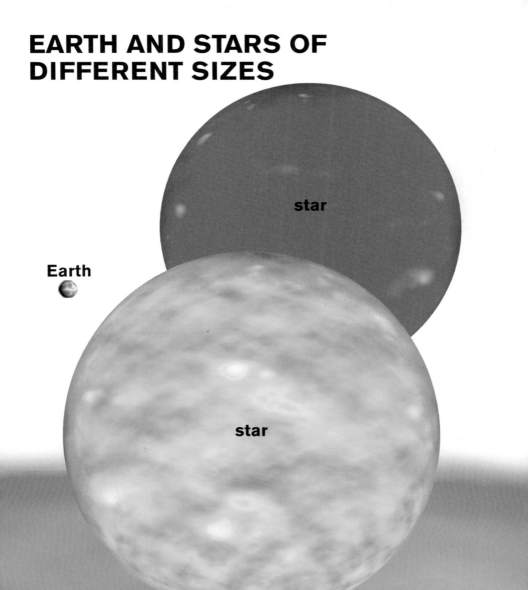

Earth

star

star

One star looks bigger and brighter than the rest. We see this star during the day. It is the Sun. The Sun is much closer to Earth than other stars. So it looks much bigger.

The Sun is a medium-sized star. Some stars are much smaller than the Sun. Other stars are much bigger than the Sun.

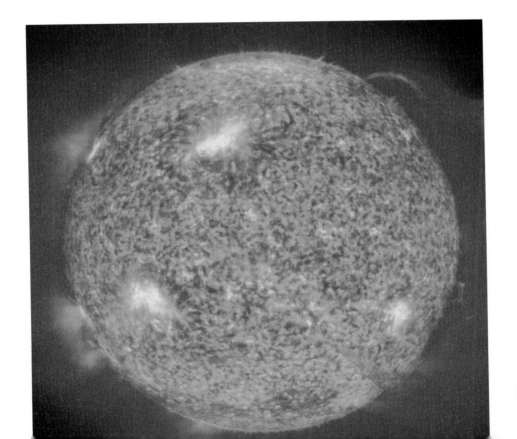

From Earth, most stars look white. But stars are not really white. They are different colors. A star's color depends on how hot it is. The hottest stars are blue. Yellow stars are not as hot as blue stars. Red stars are the coolest. The Sun is a yellow star.

New stars are always forming. Stars form inside huge clouds of gas and dust. This kind of cloud is called a **nebula.** Over time, the center of a nebula grows very hot. New stars begin to form. Then the new stars start making their own light and heat.

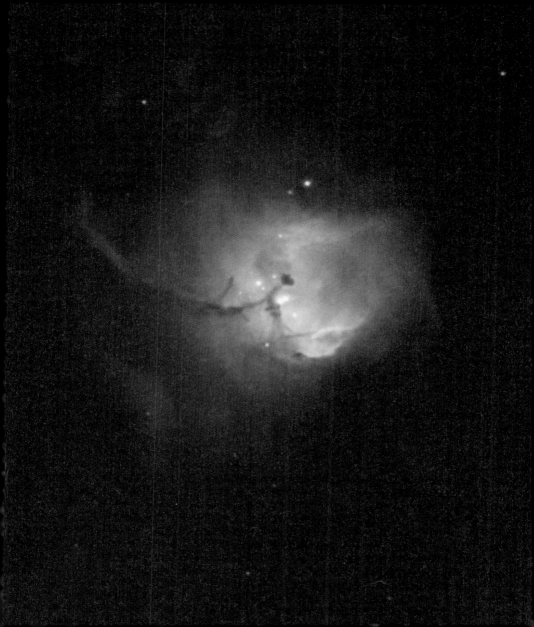

Stars do not shine forever. They die out after billions of years. Stars become cooler as they die. They swell up in size. They turn red. Then they are called **red giants.**

After millions of years, red giants begin to shrink. They become dimmer and dimmer. Finally they stop shining for good.

An artist made this picture of several red giants.

Stars that are very large die in a different way. First they become **red supergiants.** A red supergiant is as big as ten red giants.

A supergiant dies with a big explosion. The explosion is called a **supernova.** For a short time, the supernova is as bright as billions of stars put together. Then it fades away.

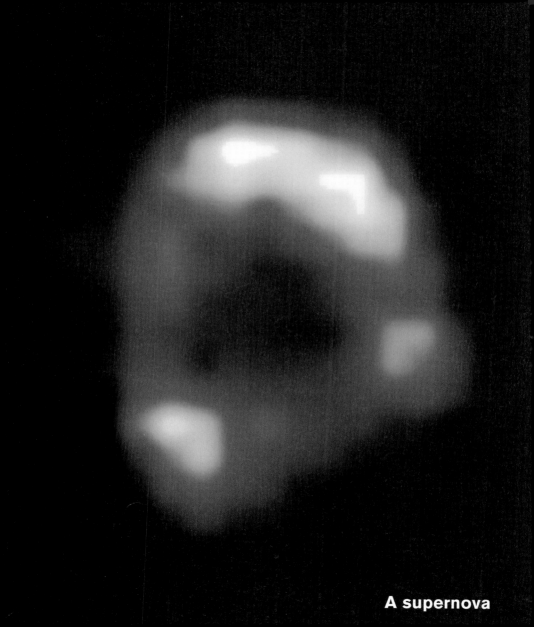

A supernova

From Earth, a person can see about 3,000 stars on a clear night. Long ago, people gave names to some groups of stars. People thought these stars looked like the shapes of people, animals, or things.

These groups of stars are called **constellations.** The Big Dipper is a famous constellation. It looks like a drinking cup with a long handle.

A **star cluster** is a large group of stars that are all in the same part of space. Some clusters are made up of millions of stars. Stars in a cluster were formed around the same time.

A **galaxy** is made of up many star clusters. Billions and billions of stars are in a galaxy. The Sun is in a galaxy called the Milky Way. The Milky Way looks like a pinwheel.

There are billions of other galaxies in space. Some galaxies look like ovals, circles, or other shapes. All of the galaxies together make up the universe.

People have always been curious about stars. But it is hard to learn about them. They are so far away. How do people study stars?

One tool people use is the telescope. Telescopes make faraway objects look bigger and closer.

The Hubble Space Telescope is one of the best telescopes ever made. It circles Earth in space. It takes pictures of outer space. The Hubble has taken pictures of faraway stars and supernovas. It has also taken pictures of nebulas and galaxies.

People also study the Sun to learn about other stars. They study the gases that make up the Sun. They study how the Sun makes heat and light.

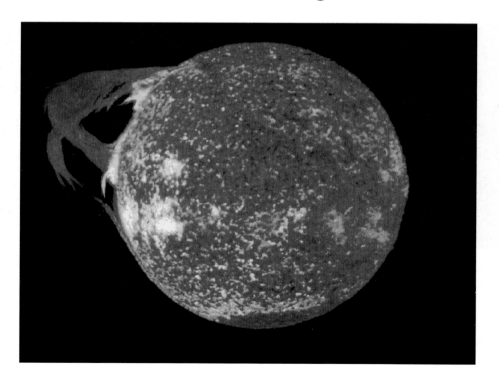

Look at the stars on a clear night. How many can you see? What would you like to know about them?

Facts about Stars

- After the Sun, the nearest star to Earth is Proxima Centauri. It is 25,000,000,000,000 miles (40,000,000,000,000 km) away from our planet.

- Sirius is the brightest star in our nighttime sky. It is 20 times brighter than the Sun. It looks small because it is so far away from Earth.

- The Sun is more than 100 times wider than Earth. Some stars are hundreds of times wider than the Sun.

- Supergiants are 1,000 times wider than the Sun.

- The smallest stars can be smaller than our planet.

- Like the Sun, some other stars have planets traveling around them.

- There may be at least 10 billion trillion stars in the universe.

- Most stars are between 1 million and 10 billion years old.

- There are 88 constellations in our sky.

- There are about 100,000,000,000 stars in our galaxy, the Milky Way.

- About 200 stars are born each year in the Milky Way.

- A shooting star is not truly a star. It is a meteor. A meteor is rock or metal from space that burns up in our sky as it falls to Earth.

- Morning stars and evening stars are not stars, either. They are planets that shine brightly at dawn and dusk.

Glossary

constellations: groups of stars that look like the shapes of people, animals, or objects

galaxy: a very large group of stars. The Sun is in the Milky Way galaxy.

nebula: a giant cloud of gases and dust where new stars are formed

red giants: small or medium-sized dying stars that have cooled and grown large and red

red supergiants: large dying stars that have cooled and grown red and very large

star cluster: a group of stars in the same part of space that are similar in age and makeup

supernova: an exploding red supergiant

Learn More about Stars

Books

Croswell, Ken. *See the Stars: Your First Guide to the Night Sky.*
Honesdale, PA: Boyds Mill Press, 2000.

Simon, Seymour. *Stars.* New York: Morrow, 1989.

Sipiera, Paul P. *Stars.* New York: Children's Press, 1997.

Websites

Kids Astronomy: Stars
<http://www.kidsastronomy.com/stars.htm>
Information about stars for kids, with helpful links to other websites about stars.

The Space Place
<http://spaceplace.jpl.nasa.gov>
An astronomy website for kids developed by the Jet Propulsion Laboratory of the National Aeronautics and Space Administration (NASA).

StarChild
<http://starchild.gsfc.nasa.gov/docs/StarChild/StarChild.html>
An online learning center for young astronomers, sponsored by NASA.

Index